MathStart

BAR GRAPHS

Lemonade for Sale

BY Stuart J. Murphy

ILLUSTRATED BY Tricia Tusa

HarperCollinsPublishers

LEVEL
3

To Harriett Barton for her support—
which has been way over the top.
—S.J.M.

To fond memories of my math teacher,
Ms. Margaret Mulvaney, who courageously wore
electric-green tights with red orthopedic shoes.
—T.T.

The illustrations in this book were done in watercolor and ink on Strathmore Bristol.

HarperCollins®, ♣®, and MathStart™ are trademarks of HarperCollins Publishers Inc.

For information address HarperCollins Children's Books,
a division of HarperCollins Publishers, 195 Broadway, New York, NY 10007,
or visit our web site at http://www.harperchildrens.com.

Bugs incorporated in the MathStart series design were painted by Jon Buller.

Lemonade for Sale
Text copyright © 1998 by Stuart J. Murphy
Illustrations copyright © 1998 by Tricia Tusa
Manufactured in China. All rights reserved.

Library of Congress Cataloging-in-Publication Data
Murphy, Stuart J.
    Lemonade for sale / by Stuart J. Murphy ; illustrated by Tricia Tusa.
        p.   cm.   (MathStart)
    "Level 2, Bar Graphs"
    Summary: The Elm Street Kids' Club decides to sell lemonade to earn money to
fix up their clubhouse and they use a graph to keep track of their sales.
    ISBN 0-06-027440-9. — ISBN 0-06-446715-5 (pbk.)
    ISBN 0-06-027441-7 (lib. bdg.)
    [1. Moneymaking projects—Fiction.   2. Graphic methods—Fiction.]   I. Tusa,
Tricia, ill.   II. Title.   III. Series.
PZ7.M9563Le   1998   96-52063
[E]–dc21   CIP
                                                               AC

Typography by Alicia Mikles
15 16 SCP 30
❖

Lemonade for Sale

The members of the Elm Street Kids' Club were feeling glum.

"Our clubhouse is falling down, and our piggybank is empty," Meg said.

"I know how we can make some money," said Matthew. "Let's sell lemonade."

Danny said, "I bet if we can sell about 30 or 40 cups each day for a week, we'll make enough money to fix our clubhouse. Let's keep track of our sales."

Sheri said, "I can make a bar graph.

I'll list the number of cups up the side like this. I'll show the days of the week along the bottom like this."

7

On Monday they set up their corner stand.
When people walked by, Petey, Meg's pet parrot,
squawked, "Lemonade for sale! Lemonade for sale!"

Matthew squeezed the lemons.

Meg mixed in some sugar.

Danny shook it up with
ice and poured it into cups.

Sheri kept track
of how many cups they sold.

Sheri announced, "We sold 30 cups today. I'll fill in the bar above Monday up to the 30 on the side."

"Not bad," said Danny.
"Not bad. Not bad," chattered Petey.

On Tuesday Petey squawked again, "Lemonade for sale! Lemonade for sale!" and more people came by.

Matthew squeezed more lemons.

Meg mixed in more sugar.

Danny shook it up with ice
and poured it into more cups.

Sheri kept track
of how many cups they sold.

Sheri shouted,
"We sold 40
cups today.
I'll fill in the bar
above Tuesday up to
the number 40.

The bars show that our sales are going up."

"Things are
looking good,"
said Meg.

"Looking good.
Looking good,"
chattered Petey.

On Wednesday Petey squawked, "Lemonade for sale!" so many times that most of the neighborhood stopped by.

Matthew squeezed even more lemons.

Meg mixed in even more sugar.

Danny shook it up with ice
and poured it into even more cups.

Sheri kept track
of how many cups they sold.

Sheri yelled, "We sold 56 cups today. I'll fill in Wednesday's bar up to a little more than halfway between 50 and 60."

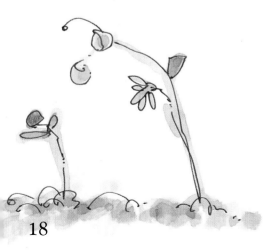

"That's great," shouted Matthew,
"That's great! That's great!" bragged Petey.

They opened again on Thursday, but something was wrong. No matter how many times Petey squawked, "Lemonade for sale!" hardly anyone stopped by.

Matthew squeezed just a few lemons.

Meg mixed in only
a couple of spoonfuls of sugar.

Danny's ice melted while he waited.

Sheri kept track
of the few cups that they sold.

Sheri said, "We sold only 24 cups today. Thursday's bar is way down low."

"There goes our clubhouse," said Danny sadly.
Petey didn't make a sound.

"I think I know what's going on," said Matthew.
"Look!" He pointed down the street.
"There's someone juggling on that corner,
and everyone's going over there to watch."

24

"Let's check it out," said Meg.

Danny asked the juggler, "Who are you?"
"I'm Jed," said the juggler.
"I just moved here."

26

Sheri had an idea.
She whispered
something to Jed.

27

On Friday, Sheri arrived with Jed.

"Jed's going to juggle right next to our stand," Sheri said.

That day Petey squawked, Jed juggled, and more people came by than ever before.

Matthew squeezed loads of lemons.

Meg mixed in tons of sugar.

Danny shook it up with lots of
ice and almost ran out of cups.

Sheri could hardly keep track
of how many cups they sold.

"We sold so many cups today that our sales are over the top.

We have enough money to rebuild our clubhouse."

"Hooray!" they all shouted. "Jed! Jed!
Will you join our club?"
"You bet!" said Jed.
"You bet! You bet!" squawked Petey.

f you would like to have more fun with the math concepts presented in *Lemonade for Sale!*, here are a few suggestions:

- Read the story with the child and describe what is going on in each picture. Talk about the graphs that accompany the story. Ask questions such as: "On which day were more cups sold, Monday or Tuesday?" and "How many cups were sold on Wednesday?"

- Talk about the different types of bar graphs that children may see. Those with bars that touch (A) or that show pictures of the items being counted (B) are often included in books that are used in schools.

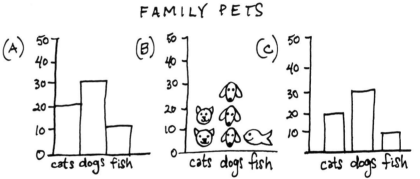

FAMILY PETS

Those with spaces between the bars (C) often appear in magazines and newspapers as well as in books. Collect examples of as many bar graphs as you can find and discuss what they mean with the child.

- Make graphs of things in the real world—children playing at the park, dogs that walk past your house, cars parked on the street, etc.—by counting them each day for a week. Do more children play at the park on Monday or Saturday? How many cars are parked on the street on Tuesday morning? How many on Sunday morning? Does the number go up or down from day to day?

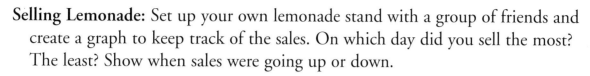

Following are some activities that will help you extend the concepts presented in *Lemonade for Sale!* into a child's everyday life.

Selling Lemonade: Set up your own lemonade stand with a group of friends and create a graph to keep track of the sales. On which day did you sell the most? The least? Show when sales were going up or down.

Around the House: Create graphs to chart how many telephone calls each family member receives or how many letters arrive each day. Who receives the most phone calls and on what days? On which days do you receive the greatest and the smallest number of letters?

Reading Books: Chart the number of books you read each week for one month. Did the number increase, decrease, or stay the same from week to week? Discuss why the number changes over a period of time.

The following stories include concepts similar to those that are presented in *Lemonade for Sale!*:

- CAPS FOR SALE by Esphyr Slobodkina

- MAMA BEAR by Chyng Feng Sun

- HOW THE SECOND GRADE GOT $8,205.50 TO VISIT THE STATUE OF LIBERTY by Nathan Zimelman